Hands on Numeracy Ages 5 – 7

Olly Octopus' Odds and Evens

odd ? even or

Odd numbers end in 1 3 5 7 or 9

Even numbers end in 0 2 4 6 or 8

0 1 2 3 4 5 6 7 8 9 10 11 12 13 14 15

What happens if you add an even number and an even number?

What happens if you add an odd number and an even number?

What happens if you add an odd number and an odd number?

Liz Webster
Linda Duncan

Acknowledgements

The authors and publishers would like to thank the children of Aldingbourne Primary School and the Vale First School for their co-operation in the making of this book. Liz Webster, Headteacher of Aldingbourne School, would like to thank her staff for all their patience, generosity and good humour over the past year. Linda Duncan would like to give special thanks to Tracy Williams and Ally Smith for their continued and positive support. Finally, they would like to thank Ian Duncan for all his hard work in helping to produce this book, and Steve Forest, the photographer, for his good humour, professionalism and endless patience.

What's the Time, Mr Wolf? (page 62)

Published by Collins, An imprint of HarperCollins*Publishers*
77 – 85 Fulham Palace Road, Hammersmith, London, W6 8JB

Browse the complete Collins catalogue at
www.collinseducation.com

© HarperCollins*Publishers* Limited 2011
Previously published in 2004 by Folens as 'Hands on Numeracy'
First published in 2004 by Belair Publications

10 9 8 7 6 5 4 3 2 1

ISBN-13 978-0-00-743933-1

British Library Cataloguing in Publication Data
A Catalogue record for this publication is available from the British Library

Every effort has been made to trace copyright holders and to obtain their permission for the use of copyright material. The authors and publishers will gladly receive any information enabling them to rectify any error or omission in subsequent editions.

Commissioning Editor: Zoë Nichols   Editor: Nancy Candlin
Page layout: Suzanne Ward   Photography: Steve Forrest
Cover design: Mount Deluxe

Printed and bound by Printing Express Limited, Hong Kong

Mixed Sources
Product group from well-managed forests and other controlled sources
www.fsc.org Cert no. SW-COC-001806
© 1996 Forest Stewardship Council

Contents

Introduction

Welcome to *Hands on Numeracy 5 – 7.*

This book aims to show how numeracy can be displayed in a creative, stimulating and fun way. We believe that numeracy can be linked to all other subject areas and does not have to be taught in isolation. By displaying numeracy in your classroom or school environment, you are highlighting its importance. Throughout this book, we aim to demonstrate how to display numeracy so that it is a valuable tool to enhance children's learning, to improve the quality of teaching and to enrich the classroom environment. Displays should always be bright, interactive and purposeful.

Numeracy should always be practical and exciting for children. This book offers a wealth of entertaining and exciting games ideas that we use in our own classrooms. We believe that using worksheets is dull, unnecessary and uninspiring. All numeracy concepts can and should be taught in a practical and stimulating way. Once the games are made, they become an invaluable resource that can be used over and over again. Children will remember games they have played rather than a worksheet they have completed!

Why Numeracy Displays are an Important Part of Your Classroom Environment

- They create an environment that is stimulating and challenging, and one in which children can reach their full potential.

- They provide an essential visual tool for learning. All the displays in this book are interactive and can be used to enhance the teaching for learning in the classroom.

- They show the importance of incorporating numeracy as part of a cross-curricular approach to teaching.

- They highlight the importance of numeracy as a fun and friendly subject to be enjoyed by all children, and not one to be afraid of!

Numeracy Wall Displays

The following ideas will ensure that your displays are stimulating, effective and purposeful.

● Use a variety of different backgrounds and borders, for example a chequerboard pattern gives an interesting and creative effect. For example, 'Dividing with Dennis the Dragon' (page 34) or 'Funny Bones' (page 56).

● Where possible, ensure that the numeracy concept is part of the title. For example, 'Pirate Pete's Place Value' (page 26) or 'Estimating Eggs' (page 60). This will give the display an instant focus and visual reference.

● Be original and daring with your titles! Using alliteration is a fun and child-friendly tool. It instantly captures their imagination and is an easy way for children to remember a particular numeracy concept. For example, 'Round Up Rocket' (page 22) or 'Captain Capacity' (page 68).

● Include the mathematical vocabulary on each display. It must be BIG, BOLD and BEAUTIFUL, not little, lifeless and lost on the wall.

● Using questions on the display ensures that children use it to enhance their learning. This, in turn, makes it a teaching aid. For example, 'Flying Frogs' (page 8) or 'Handa's Data Handling' (page 46).

Numeracy Table-top Displays

● These are a valuable resource and allow the children the opportunity to interact with the learning environment in a practical way.

● They provide an instant link to the wall display, ensuring that the children are accessing their kinaesthetic learning as well as their visual learning. For example, 'Sir Symmetry' (page 54) in which the children look at the wall display and reinforce their learning by using and carrying out the activities on the table-top display.

● The table-top displays should be an extension of the wall display and, therefore, must be as equally bold, bright and interesting and easily accessible to children. They are there to be used and not just to look beautiful!

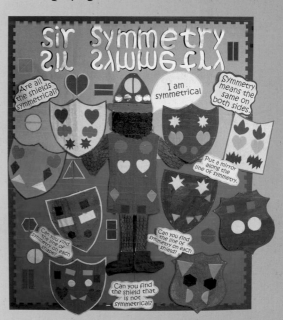

When teaching numeracy remember:

Needs to be fun

Use a variety of teaching styles

Make it memorable

Exciting

Rewarding

Always be active

Challenge the children

You can make a difference!

Liz Webster and Linda Duncan

Counting Clowns

Oral and Mental Warm Up

- Prepare ten juggling balls made from cardboard circles and number them one to ten, in both words and numerals. Put them on a table or on the floor. With you in role as Clever Clown, explain to the class that the juggling balls have got mixed up and that you need their help to sort them into the correct order.

- Ask pupils to sit in a circle and count to ten, clap ten times, get up and jump ten times, and so on.

- Make a set of one to ten numeral cards, a corresponding set of number word cards and a set of domino cards. Shuffle all the cards together and give one to each child. On your signal (a bell, for example), challenge the children to find their matching numeral, domino and number word, and to get into a set of three. Shuffle the cards and play again.

Focus of Learning

Counting and recognising numerals from one to ten

Art and Display

- Use paint and collage materials to create giant clowns for the display.

- Paint giant bow ties. Give the children a number and ask them to paint or add collage materials showing the number as spots, strips or squares.

- Draw and colour pictures of clowns.

- Make clay bow ties.

- Make funny clown hats. Children pick a number up to ten out of a bag and stick the correct number of spots on to their hats.

Practical Activities

- Play 'Spin and Match'. Ask the children to sit in a circle and place an empty bottle in the middle. Place the number word, numeral and domino cards (see 'Oral and Mental Warm Up', page 6) inside the circle. Spin the bottle. Whichever child the bottle points to picks up three corresponding cards.

- Play 'Number Lotto'. Make a set of clown lotto gameboards with the numbers 1 to 10 on balloons or juggling balls. Pupils take turns to pick a number word out of a box or bag. He or she then covers the corresponding number on their lotto board with a counter.

- Play 'Musical Counting'. With the class sitting in a circle, hand out a set of 1 to 10 numeral cards. When the music starts, pupils pass the numbers around the circle. When the music stops, the children holding the numbers stand up one by one, in the correct order, and shout their number. This can also be played with numeral words.

- Play 'Make a Model Number in One Minute'. Give each child a ball of modelling clay and a number word card. Using a sand timer, explain that they have to make the number that is written on the card before the timer runs out.

7

Flying Frogs

Oral and Mental Warm Up

Focus of Learning

Using a number line

- With the class, practise counting to ten forwards and backwards.

- Play 'Line Up the Lily Pads'. Make a set of large lily pads numbered 0 to 10. Each child takes it in turn to pick up a lily pad. The object is to order the lily pads correctly. Ask a child wearing a frog mask or hat to jump from 0 to 10 and back in ones while the rest of the children count aloud.

- Use the number line from the above activity. Ask one 'frog' to stand on a number, for example 3. Pose the question, "How many jumps will the frog make to get to number 6?" Ask the 'frog' to test the class's answers by jumping to see who is right.

- Discuss the concept of adding using the number line. Give a 'frog' on the number line a card with an addition calculation on it, for example 1 + 2. Ask the frog to use the number line to find the answer.

Art and Display

- Look at a picture of Monet's *Water Lilies*. Use chalk pastels to create a large Monet-style water lily picture as a display background. Paint large frogs and use paper collage to make ten large lily pads and stick these to the background. Add numerals to the lily pads.

- Make frog masks and hats using card and collage materials.

Practical Activities

- Make a set of 0 to 10 lily pad counting lines using green circles stuck onto a card strip. Prepare a die with +1, +3, +2 on three faces and a fly on the other faces. Give each child a counter to move up the lily pad line. Take it in turns to roll the die and move the counter up the lily pads. If the children roll a fly, the frog is having lunch so they miss a go!

- Play 'Fishing for Frogs'. Each child makes a 0 to 20 lily pad number line. Provide a set of numeral frog cards with the numbers 0 to 10. Attach paperclips to the frog cards and place in a container. Using magnetic 'fishing' rods, each player takes it in turn to fish out two frogs. They find the bigger number on the lily pad number line and then add on the second number by 'hopping' the smaller number along the line.

9

Bob Builds in Threes

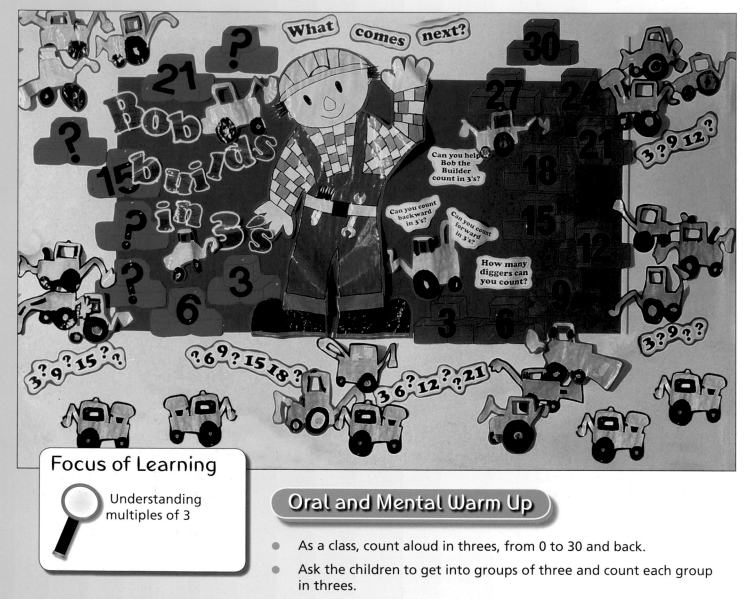

Focus of Learning

Understanding multiples of 3

Oral and Mental Warm Up

- As a class, count aloud in threes, from 0 to 30 and back.

- Ask the children to get into groups of three and count each group in threes.

- Seat children in a circle. Count around the circle beginning at number one. Child number three stands up, child number six stands up, and so on. (This activity can be played with other multiples.)

- Make a set of number cards with 3, 6, 9 … 30. Give each child a card and challenge them to get into the correct order.

- Cover the multiples of 3 on a class 100-square. Ask children to fill in the missing numbers.

- Play 'I Spy the Wrong Number'. On the board, write a selection of multiples of 3 up to 30. Include another number, such as 5, 11 or 2. Encourage the children to spot the wrong number and to give reasons for their choice.

- Give each child a whiteboard. Start counting up from 3 for five or six multiples but leave a number out, for example, 3, 6, ?, 12, 15. The children have to write the missing number. Repeat, but starting from a different multiple of 3.

Art and Display

- Paint and use collage materials to make a large picture of Bob the Builder. Draw and paint diggers for the display.

  - Use binca material and thread to sew a picture of a house.

  - Use clay to make a 3D house.

Practical Activities

- Give each child a numbered 100-square and ask them to colour in the multiples of 3. Then hand out blank 100-squares and ask the children to colour the multiples of 3 squares. What pattern do they notice? Then ask the children to write the multiples of 3 on the appropriate coloured-in squares.

- Use giant bricks and ask the children to make sets of three bricks. Draw and label the sets of bricks by counting in threes.

- Put the children in groups and make a set of multiples of 3 cards, from 3 to 30, for each child in the group. Place the cards face down on the table. The object of the game is to pick the cards in the correct order and so the children take turns to pick a card. For example, child one picks a 6 and so she puts the card back. Child two picks a 3 and so he keeps the card, and so on until one of the group has collected cards to 30.

- Turn the role-play area into a building site. Make a set of giant multiple of 3 cards from 3 to 30. Hide the cards around the building site. Ask the children to find the cards in the correct order and to record where they found each card. For example, '3 is under the bricks', '6 is in the cement mixer', '9 is in a sack'.

- Number a set of hard hats with multiples of 3 from 3 to 30. Give each child in the group a hard hat and ask them to get into the correct order.

- Using the hats from the above activity, put a hat on each child's head without telling them their number. The children have to use their knowledge of counting in threes and their awareness of other children's numbers to get into the correct order.

Olly Octopus

Focus of Learning

Recognising odd and even numbers

Oral and Mental Warm Up

- Tell the children that Olly Octopus is going to visit (he could be you in role, a puppet or a child dressed as an octopus). Olly has some odd numbers written on his tentacles. Ask the class what they notice about his numbers. Explain that they are all odd numbers and talk about the concept of an odd number.

- Say that Olly is sad because he has lost some of his odd numbers. Place a selection of odd and even numeral cards around the room. Invite the children to help Olly to find all the odd numbers.

- Teach Olly's 'Simple Secret'. Explain that all numbers that end in 1, 3, 5, 7 and 9 are odd numbers and all numbers that end in 2, 4, 6, 8 and 0 are even numbers. Now test the children. Write a BIG number on the board, such as 3451, and ask pupils to decide whether it is odd or even, and to give their reason.

Art and Display

- Create a large octopus using paint and collage materials and print a sea background using dark blue paper overprinted with white circles to represent bubbles. Sponge-print the bottom of the sea in shades of yellow and orange. Paint and use collage materials to create a seaweed effect and individual fish.

- Use a torn-paper technique to make fish for an underwater mobile.

- Use a marbling technique to create a watery effect for a seascape background and use oil pastels or collage materials to make fish. Dribble PVA glue onto white calico in the shape of a fish (or any underwater creature) and allow it to dry. Use fabric dye to paint. When dry, peel off the PVA glue for a stunning effect!

Practical Activities

- Make a set of fish-shaped cards with the numbers 1 to 20 (larger numbers can be used). Then make a set of octopus cards, some with the word 'odd' and some with 'even' written in bold. Give each child a card (either a fish or an octopus) and, on a signal (a bell, for example), ask the children to partner up a fish and an octopus correctly.

- Play 'Organise Olly's Odds and Evens'. Make a set of gameboards divided into two. On one side draw a large octopus with the word 'odd' written on. On the other side draw a large fish and write the word 'even'. Using 0 to 9 numeral cards, the children work in pairs and pick two or three cards. They decide what 2- or 3-digit number it could make and write it on the octopus or the fish. Encourage the children to rearrange the numeral cards to make different numbers.

- Ask the children to find out what happens when you add an odd number and an odd number, an even number and an odd number, an even number and an even number.

- Play 'What's My Number?' Make a set of bingo-style gameboards with different amounts of 'odd' and 'even' written on each one. Using 2-digit numeral cards, pick two cards and invite the class to predict whether the cards' total would be odd or even. For example, 93 + 21 = even. The children cover the correct answer on their bingo board. The first to cover all the words on their board is the winner.

13

Little Robin Red Vest

Focus of Learning

Using ordinal numbers

Oral and Mental Warm Up

- Ask the children what the first thing was that they did this morning. What was the second thing they did?

- Ask the children to say what they know about ordinal numbers and when they are used. If they were in a race, would numbers like 1 to 7 be used for the finishing positions? Why not? What would be used?

- Show the children vests with 1st, 2nd, 3rd up to 7th written on. Hand out the vests and ask the children to order themselves correctly. Take away a vest. Which one is missing? Then repeat the activity using labels with the words 'first', 'second', 'third' … 'seventh' written on.

- Play a matching game. Give out both the vests and the labels. Play some music and ask the children to find their partner when the music stops.

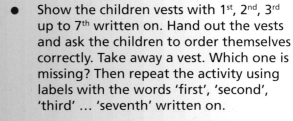

Art and Display

- Paint patterned vests for Little Robin. Use the colours from the story, Little Robin Red Vest by Jan Fearnley (published by Mammoth, 1999).

- Paint a large robin for the display.

- Make small robins using circles, semicircles and triangles. Label them 1st, 2nd 3rd to 7th and use for a table-top ordering activity.

- Make clay robins and label them one to ten. Use as an ordering activity by matching each robin to the correct vest.

- Make a set of coloured vests with ordinal numbers on and a set of large robins with ordinal words on. Laminate and use for a variety of games listed below.

Practical Activities

- Make a board game using coloured vests from the story to match the ordinal word with the ordinal number. Ask the children to spin the spinner, which is marked with ordinal numbers, and then cover the corresponding vest on the board, which is marked with ordinal words.

- Make a game using small coloured vests labelled 1st, 2nd, 3rd … 7th. You will need one set for each child playing the game. Put the vests in a box in the middle of the table. Ask the children to take it in turns to pick out a vest. The object of the game is to be the first to collect a complete set of vests in the correct order. Several small robins could be added to the box as jokers. If a child picks out a robin, they lose all their vests!

- Label a set of coloured bibs with the ordinal numbers 1st to 10th. Make a set of large robins with the ordinal words 1–10 and place them around the hall or outside area. Give each child a vest to wear. On a given signal, ask the children to find their corresponding robin. Repeat this activity several times by changing vests and position of robins.

- Try the 'Find the Robin' game. Ask the children to sit in a circle. Place the vests and clay robins labelled 1–10 (see 'Art and Display') in the middle of the circle. Spin the bottle. Wherever the bottle stops, that child picks up a robin and its matching vest.

- Use a programmable robot and make it into a robin. Place the vests made in 'Art and Display' in order around a large space. Ask the children to program the robin to visit the vests in the correct order.

Counting Blue Balloons

Focus of Learning

🔍 Ordering numbers to 20

Oral and Mental Warm Up

- Count from 0 to 20 and back with the class. Then repeat using a different starting number each time.

- Play 'Guess the Missing Number'. Count up to 20 but miss out a number. The children have to guess the number.

- Give each child a 1–20 number fan. Ask questions such as, "What number comes between 7 and 9?" "Show me a number that is more than 15." Children use their fans to show the answer.

- Make a human number line. Give each child a number card and ask them to get in the correct order.

- Using the human number line, make a variety of statements such as, "Sit down if your number is more than 18." "Less than 6." "Between 10 and 12."

- Draw a blank number line on the board labelled 0 at one end and 20 at the other end. Give out a selection of cards between 1 and 20. Ask the children to use sticky tack to stick the cards on the number line, and to give reasons for their choice.

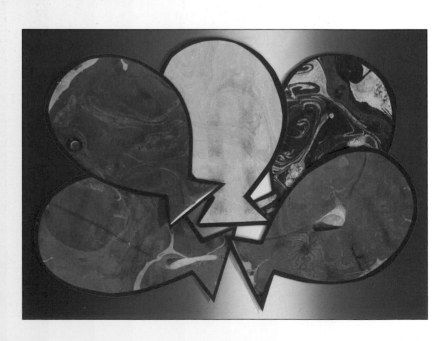

Art and Display

- Weave some 2D balloons using different shades of blue paper.

- Create some colourful blue balloons using marbling.

- Make clay balloon-shaped tiles and use tools to add texture and pattern. Spray the tiles blue.

- Make a giant blue balloon using a variety of collage materials.

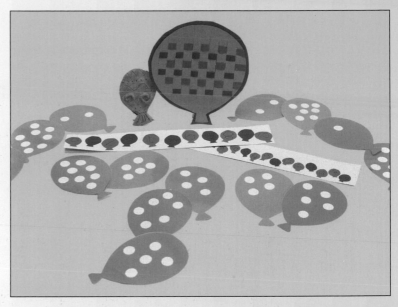

Practical Activities

- Prepare some blue balloons with different numbers of dots on, up to 20. Place the balloons around the classroom. Ask each child to find a blue balloon, count the dots and write the corresponding numeral and word on their whiteboard. For example, if they spot a balloon with three dots, they write: 3, three.

- Make a blue balloon number line. Each child cuts out 21 blue circles and writes the numbers 0 to 20 on each circle. The children arrange the circles on a long piece of card to make a number line.

- Hang balloons numbered from 0 to 20 from a washing line. Burst some of the balloons. Ask the children which number balloons have burst.

- Ask the children to sit in a circle. Tell them who is number 1 (the first in the circle) and who is number 30 (the last in the circle). Ask them to stand up if they think they are number 5, 8, 23, and so on.

- Play 'Balloon Bingo'. Give each child a balloon gameboard with some numbers up to 10 on. Pick balloon numbers out of a box. Children tick any numbers that match on their gameboard.

The Jolly Jigsaw

Oral and Mental Warm Up

- Look at a large 100-square. With the class, count from 0 to 100 in ones, twos, fives and tens.

- Play 'Find the Missing Number'. Remove or cover a number. Ask the class what number is missing and to explain why.

- Give pupils a large section from a 100-square, cut out like a piece of a jigsaw puzzle. Ask the children to put the 100-square puzzle back together.

Focus of Learning

Using and understanding a 100-square

- Play 'Guess My Number'. Tell the children you are thinking of a number. Give them a clue such as, "It is more than 27". Using a class 100-square, one child removes or covers all the numbers below 27. The next clue could be, "It is an even number between 60 and 96" and a second child removes or covers all the numbers below 60 and above 96. A third child could conceal all the odd numbers. Give clues until the children have guessed the number correctly.

Art and Display

- Design and paint a pattern or picture to make individual jigsaw puzzles.

- Cut out an individual jigsaw piece to use as a template. Children draw around the template and colour their jigsaw piece.

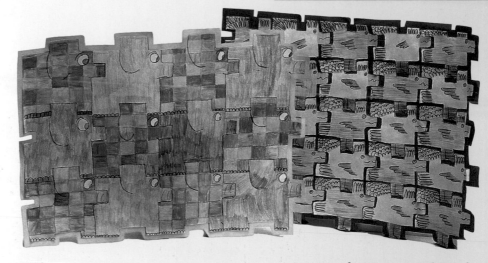

Practical Activities

- Cut a set of 100-squares into six jigsaw pieces. On the back of each piece, write a number from 1–6. Each child then takes it in turn to roll a die, pick up the corresponding piece of puzzle and begin to make a jigsaw. The first person to complete their 100-square is the winner.

- Play 'Hoot the Human 100-square'. Using a large 100-square mat, give each child a problem such as, "What is the total of 8 and 7?" "What is the sum of 19 and 14?" "What is the difference between 26 and 32?" At the sound of a hooter, the children jump to the correct square!

- Hand each child a 100-square. Make a set of cards with instructions such as 'Cover six "noughty" numbers' (numbers ending in zero) or 'cover 12 even numbers', and so on. The children take it in turns to pick a card and carry out the instruction.

- Give each child a blank 100-square and play the same game. This time they write in the correct numbers.

- Hand out portions of a 100-square with numbers missing to each child and ask them to fill in the missing numbers.

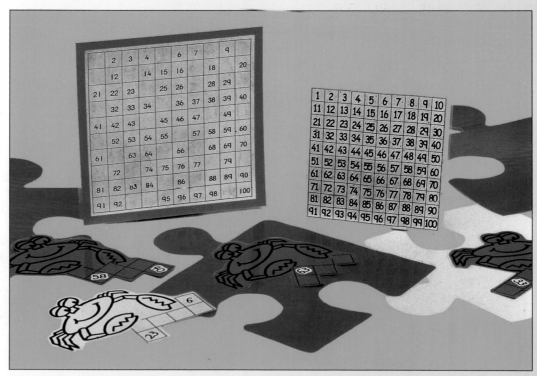

Perfect Pairs

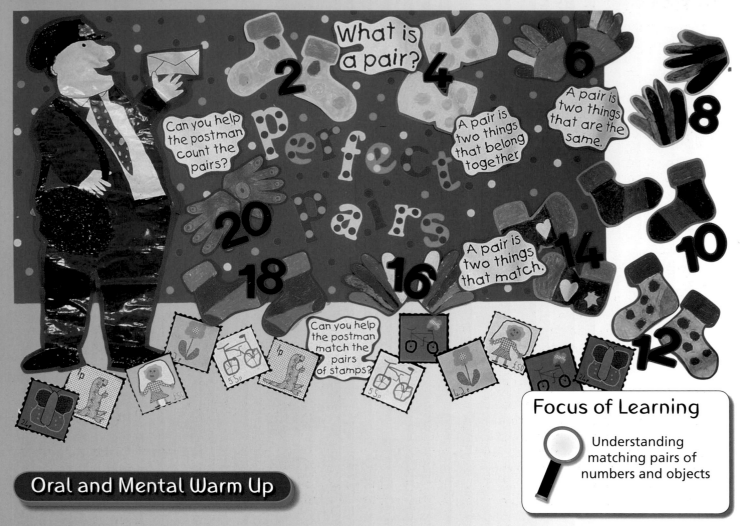

Focus of Learning

Understanding matching pairs of numbers and objects

Oral and Mental Warm Up

- Ask the children to take off their shoes and put them in the middle of the carpet so that they are all muddled up. Ask each child to find a matching pair of shoes. Discuss what 'a pair' means.

- Play the 'Spin a Perfect Pair' game. Put a selection of pairs of shoes, gloves and socks in the middle of the circle and spin the bottle. Whichever child the bottle points to has to find a perfect pair. When a pair is found, the rest of the children shout, "Perfect pair!"

- Wrap pairs of empty boxes in patterned wrapping paper and put them in a large sack. With you in role as the Jolly Postman, ask the children to help sort the parcels into matching pairs.

- Place two sets of 1 to 10 numerals around the classroom. Ask pupils to find the pairs of numbers and sticky tack them to the board. Prompt individuals to add each pair of numbers to find the total, for example, $4 + 4 = 8$.

- Repeat the above activity, but this time the object of the game is to say which 2-digit number the children have made by collecting a perfect pair, for example, 44 is forty-four.

Art and Display

- Paint and use collage materials to make a giant postal worker.
- Use oil pastels to design pairs of patterns on sock-, glove- and boot-shaped paper.
- Use ICT to create some matching pairs of postage stamps.

Practical Activities

- Ask the children to find a partner. The pairs work together to paint identical patterns on glove-, sock- or boot-shaped paper or paint a perfect pair of numbers.

- Play 'Perfect Pairs'. Make a set of character pair cards, for example Goldilocks on one card and the Three Bears on another, or the Wicked Witch and Hansel and Gretel. Then play Pelmanism. Alternatively, you could use pairs of number cards or object cards.

- Play 'Pounce on a Perfect Pair'. Place a large selection of pairs of boots, socks and gloves at random around the room. Prepare a large die with the letter P on three of the faces and a star on the other three. Each child

sits in a hoop. The teacher rolls the die. If it lands on the star the children pick any item. If it lands on a P they pick a Perfect Pair! The child who collects the most pairs wins the game.

- Play 'Perfect Pair Bingo'. Make a set of bingo gameboards with random selections of the numbers 2, 4, 6, 8, 10, 12, 14, 16, 18 and 20 and have ready two sets of number cards 1 to 10. Pick up two matching cards and ask the children to add them together. When they have worked the pair out, they try to find that number on their gameboard and cover it with a counter. The first child to cover all the numbers on his or her board wins the game.

Round Up Rocket

Oral and Mental Warm Up

- With the class, count in tens from 10 to 100 and back. Repeat, using different multiples of 10 as a starting point.

- Make a human number line using multiples of 10 cards up to 100. Ask the children to get into the correct order.

- Create hats with a giant star on the front and a multiple of 10 up to 100 written on for each child. Hand out rocket-shaped cards with random numbers to 100 (you will need enough cards for one per child). Ask the children to find their nearest multiple of 10 and to give reasons for their choice.

- Play 'Round Up Rocket'. Take the children into a spacious area or outside. Prepare planet-shaped cards showing multiples of 10 up to 100 and place them around the area. Put the children into teams of astronauts, giving them a hoop as their spaceship. Then give each team a card with a sum on, such as 27 + 41. The astronauts work out the answer to the nearest 10, and one member of the team 'launches' him or herself to the correct planet shouting, "Round up Rocket". The first to the correct planet wins that number of points for his or her team. For example 27 + 41 = 68. The nearest multiple is 70, so they collect 70 points. The object of the game is to reach 1000 points exactly!

Focus of Learning

Understanding how to round a number up or down to the nearest ten

Art and Display

- Paint giant rockets for a wall display. Use watercolour pencils on paper to make astronauts.

- Make starry pictures by sponge printing a background in shades of blue. Overprint stars in a deeper shade of yellow or silver.

- Look at a picture of *Starry Night* by Vincent Van Gogh. Use thick paint and palette knives to create pictures in the same style.

Practical Activities

- Play 'Round Up Bingo'. Make a set of planet grids with a multiple of 10 written on each planet. Roll two dice or pick two number cards to make a 2-digit number. The children cover the nearest 10 with a rocket-shaped counter or piece of card.

- Play 'Hop to the Hoop'. Place ten hoops on the floor, each with a number card showing a multiple of 10. Ask the children to sit in a circle. Give out random number cards or sum cards. On a given signal, pupils hop into the correct hoop.

- Play 'Zoom to the Moon'. Make a set of cards, each with the multiples of 10 to 100 written in separate moons and a rocket at the bottom. Prepare a separate

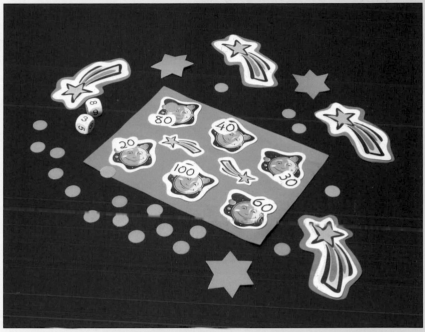

set of cards with additions on, such as 27 – 9 = 18. Each child takes an addition card, writes the answer on a rocket and draws a line from the answer to zoom to the correct moon to the nearest ten.

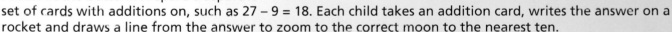

- Play 'Musical Round Up'. Make a set of rocket cards with near tens numbers between 1 and 100, such as 18, 39, 48, 57, etc. To the tune of *London Bridge is Falling Down*, sing, "Pass the numbers round and round, round and round … stop right now. Who has got a number rocket, number rocket … Stand up now". The children with that card stand up and say their nearest 10.

Fruity Fractions

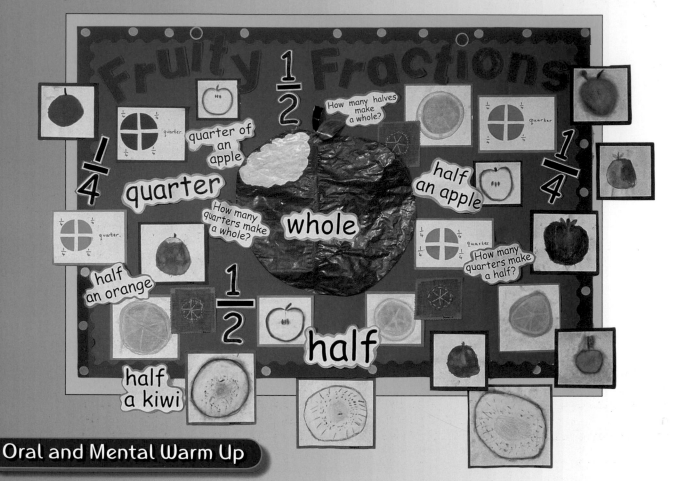

Oral and Mental Warm Up

- Show the children a whole apple. Discuss the concept of 'whole'. Then cut the apple in half and discuss the concept of 'halving'. Cut the apple into quarters and talk about the concept of 'quarters'.

- Give each child an apple-shaped piece of paper. Ask them to fold the paper in half. Open up the paper and discuss how many halves make a whole. Then ask the children to fold their paper apple into quarters. Open up the paper and discuss how many quarters make a whole, and how many quarters make a half.

- On the board, write the symbols for halves and quarters and show how to write the words.

- Make a set of 'fruity fraction' picture cards and ask the children to match the words and symbols to the pictures.

- Play 'Feel the Fraction'. Prepare a container containing whole, halves and quarters of real fruits, and give each child a whiteboard and pen. Ask individuals to pick out one of the fruits and show it to the rest of the class. The children have to write down the appropriate fraction symbol and word.

Focus of Learning

Understanding simple fractions

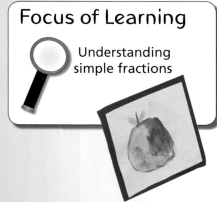

Art and Display

- Paint a large apple for the wall display and use pastels to draw apples.

- Make observational drawings. Use pastels to draw half an orange, half a kiwi fruit and half an apple.

- Sew half a lemon, or another fruit, onto hessian.

Practical Activities

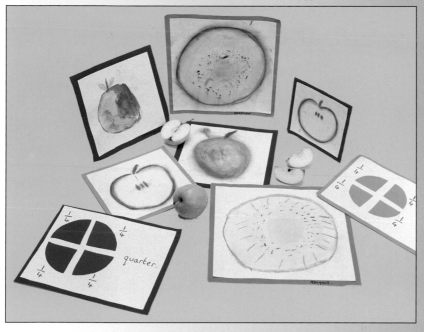

- Give each child coloured paper circles and contrasting paper. Ask them to cut the circles in half and in quarters, and to stick these onto the contrast paper. Invite pupils to write the appropriate fraction symbol and word.

- Spin a Fruity Fraction! Make a set of gameboards with four or more apples drawn on. Make a spinner with $\frac{1}{4}$, $\frac{1}{2}$, 1 and a star on it. Children spin the spinner and colour in the appropriate section of apple. The star indicates miss a go. The winner of the game is the one who colours in all of the apples. They must spin the correct fraction to colour in. For example, if a child has a quarter of an apple left to colour, he or she must spin a quarter.

- Ask children to divide larger numbers into halves and quarters. They might like to use cubes to help, for example by sharing 20 cubes between four.

- Play 'Human Fruity Fraction Puzzles'.
  Give each child a section from a giant paper apple. When the hooter sounds, they have to make a whole apple by joining up with other children. For example, two children who each have half an apple would get together and make their apple on the floor. Make it more difficult by asking pupils to make their puzzles in different ways, such as: one half and two quarters. Players could sit in a hoop when they have made a whole apple!

- Play 'I Spy Fruity Fractions'. Place fruity fraction pictures around the room. Provide the children with clues such as, "What fruity fraction is near the cupboard?" When pupils have found it, they write the symbol and word on their whiteboards. The class can check their answer and find the next clue on the back of the fruity fraction picture.

Pirate Pete's Place Value

Focus of Learning

Understanding place value

- Tell the class that Pirate Pete (you, in role) is going to visit the classroom to find out who has stolen his gold! He has sacks containing different amounts of gold. For example, one sack has 20 pieces, another has four. Ask the children how much 20 and 4 makes. Talk about place value and ask them the value of the 2 in 24 and the value of the 4. Continue with different numbers.

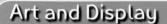

- Show the children three sacks. If one sack has 100 pieces, another has 30 pieces and the third has seven pieces, how much is it altogether? What is the value of the 1 in 137?

- Play 'Ship Ahoy'. Make a set of 2-digit number sacks, such as 34, 56, 21, and a set of cards with the corresponding numbers such as 3 tens, 4 units, 5 tens, 6 units, 2 tens, 1 unit. Give out the sacks and cards and, when you shout "Ship ahoy", the children get into the correct groups of three. This game can also be played with 3-digit numbers with children getting into groups of four.

Art and Display

- Make a large pirate for display using paint and collage materials, and paint large sacks.

- Draw individual pirate pictures.

- Use a paint package on the computer to make a pirate picture.

- Make pirate masks from collage materials.

Practical Activities

- Play 'I Spy the Booty'. Place large sacks of booty labelled with 2- or 3-digit numbers around the classroom. Ask the children to visit each sack and record the number of tens and units (or hundreds, tens and units) on their whiteboards.

- Make a set of 'booty gameboards' with sacks labelled with 2- or 3-digit numbers, and prepare gold pieces with the values 100, 10, 1, 2, 3, 4 and 5. (You will need a lot of these!) Give each child a gameboard. Take it in turns to pick a number card and place the number in the correct place on the sack. The object of the game is to collect the correct value for each sack of booty.

- Play 'I've Got the Booty' (Pelmanism). Make two sets of booty cards, one set with the number in digits and the other with the number in words. The children take it in turns to turn over two cards and try to make a matching pair.

- Give each child three number cards. On the sound of a bell, the children make the biggest number and the smallest number they can. Discuss the numbers they have made and then ask them to order all the class numbers from the smallest to the biggest. An extension activity could be to ask the children to find the difference between their two numbers.

- Play 'Bag the Booty'. Make a set of large booty sacks, one for each child in the class. Also prepare a set of smaller, colourful booty sacks, some labelled with 10, some with 1-digit numbers and, for extension work, some labelled in hundreds. Sticky tack a random selection of booty sacks to the board, such as 10, 10, 10, 7 and 2. The children write the total on their sack and shout, "Bag the booty!" when they have completed the sum.

Light Up the Lighthouse

Focus of Learning

Understanding the different vocabulary for addition and practising counting on

Oral and Mental Warm Up

- Make a large lighthouse out of cardboard. On it write a selection of addition problems such as 'the total of 4 and 12', '17 plus 22' or 'increase 45 by 9'. Write the answers on 'light rays' and put them in a bag. Tell the class that Larry the lighthouse keeper has a problem. He cannot add up the numbers on his lighthouse to make the light flash and he needs the children's help. As the children give the correct answer to each calculation, you, in role as Larry, add the corresponding light ray and light up his lighthouse.

Art and Display

- Print the sky and sea to make a background for the main display.

- Use wax crayons to draw lighthouses. Then colour wash them with thin paint.

- Use recyclable materials to make individual lighthouses.

Practical Activities

- Play 'Boats and Lighthouses'. Make a set of lighthouse cards with a number on the back and prepare a set of boats with an addition sum on the back. Take it in turns to pick up a lighthouse and a boat. If they match, they keep the cards. The child with the most cards at the end is the winner.

- Play 'Leg it to the Lighthouse'. Place a large blank lighthouse on the wall in the hall or outside area. Sit each child in a hoop (a boat) and give each one a red or white piece of card with an addition sum written on it. Put a number on the top of the blank lighthouse. The children with the corresponding sum cards 'Leg it to the Lighthouse' when the hooter hoots and stick the sum card to the lighthouse.

- Make a set of six lighthouses with a number on the light. Prepare a set of large digit cards numbered 1 to 20 – enough for one per child. Give six children a lighthouse and the rest of the class a digit card. Ask the children holding the lighthouse cards to light up their lighthouse by finding three numbers to total the number at the top. The object is to be the first group to light up their lighthouse and make the foghorn sound (a hooter).

- Play 'Last to Light the Lighthouse'. Make a set of bingo lighthouse cards with addition sums on each lighthouse. Call out a number and the children with the correct answer cover the correct lighthouse with a counter. The object is to be the last to cover the lighthouses!

Subtracting Sunflowers

Oral and Mental Warm Up

- Put a variety of operation words on the board, for example 'take away', 'subtract', 'minus', 'add', 'more than', 'less than', 'plus' and 'total'. Ask the children to sort out the take away words. Discuss the meaning of 'take away'.

- Make a large sunflower with removable petals. Ask the children to count the petals first, then remove five. How many petals are left? Continue this activity, taking away different numbers of petals each time. Use a wide variety of number language so that the children become aware of the different ways of saying 'take away'.

Art and Display

- Make a large sunflower for the centre of the display. Create observational drawings of sunflowers using sketching pencils. Use chalk pastels to draw sunflowers. Use torn paper technique to make sunflowers.

- Create a textured paint by adding salt to thick yellow poster paint. Use palette knives to create sunflowers.

- Make sunflower masks/headpieces.

Practical Activities

- Play 'Sort Out the Sunflower'. Make a set of circles to represent the centre of the sunflower, each with a different number on and enough for one per child. Also prepare a set of petal cards, each with a subtraction sum written on. There should be ten petal subtractions for each sunflower centre answer. Give each child a circle. Put the petal cards in a basket in the middle of the table. The children take it in turns to pick a petal card and work out the subtraction sum. If the sum corresponds to the number in the middle of their sunflower centre, they keep the petal. The object of the game is to collect ten petals and make a sunflower.

- Play 'Spot the Sunflower'. Make a set of large sunflowers, each with a number written on. Place these around an outside area, hall or classroom. Give each child a subtraction sum, for example '18 minus 3', 'decrease 12 by 2', '4 less than 15', and so on. On a given signal, the children find the correct sunflower.

- Use the sunflower cards from 'Spot the Sunflower', but this time give each child a whiteboard and pen. Ask the children to visit each sunflower and write a different type of subtraction sum.

- Play 'Write a Sunflower Story'. Give the children a subtraction sum such as '15 minus 10'. Ask them to write their own number story to share with the rest of the group. For example, 'There are 15 sunflowers smiling in a field. The farmer comes along and picks 10 for his wife's birthday. There are 5 sunflowers left.'

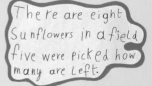

You go to the shop to buy some sunflowers. There are twenty sunflowers and you buy five, how many are left?

There are eight sunflowers in a field five were picked how many are left.

Daisy, Daisy

Oral and Mental Warm Up

- Make 0 to 10 daisy stepping stones and display them on the board. Give one of the Crazy Daisy petals to a child and ask him or her to make 10, using the number line for help. The child writes the appropriate number on Crazy Daisy's petal. Continue until all the petals have been completed.

- Sing an alternative version of the traditional song, *Daisy, Daisy*.

  Daisy, Daisy, ten is the answer for you,
  I'm half missing
  Give me the answer, do!
  I don't want to make it too hard
  So I'll give you each your own card
  And you'll look neat
  When you both meet
  And the answer you make is ten!

Make a set of daisy cards each with a number 0–10. Give half the class a card each. Sing the song and children with the cards have to find their partner before the song finishes!

Focus of Learning

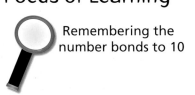

Remembering the number bonds to 10

32

Art and Display

● Make individual daisy pictures using a variety of collage materials. Use a different texture for each petal.

● Use oil pastels to draw daisies on calico, then colour wash them with thin blue paint.

● Use chalk pastels to draw daisies.

● Use block printing to print a repeating daisy pattern.

Practical Activities

● Make a set of giant daisies and write one number from 0 to 10 on each petal. The children take it in turns to spin the spinner or roll a die and write the corresponding number on the correct petal. The object of the game is to complete all of the petals with number bonds to 10.

● Make a set of daisy gameboards – one per child. On each gameboard draw several daisies, each with a different number of petals (up to 9). Write the number 10 in the middle of each daisy. Invite the children to draw on the correct number of petals.

● Play 'Pick a Daisy'. Make two sets of 0 to10 daisy digit cards. Ask each child in the group to take it in turns to turn over two cards. The object of the game is to make 10. To make the game more exciting, add a few bumble bee cards. If one of these is picked the child has to put all his or her cards back!

● Play 'Daisy, Daisy'. Reuse the large daisy stepping stones from page 32, adding the corresponding number bond to the back of each card. Place each with the single number uppermost, on the carpet with the group of children sitting around. Ask one child to stand on the first daisy, say the number and then choose another child to guess the number bond on the other side. Turn the daisy over to check. If correct, the child moves to the next daisy, and so on. The object is to get to the end of the daisy chain … and back!

● Make 11 big petals and write the numbers 0 to 10 on each petal. Attach them round your waist. Tell the class that you have a little problem. The Duchess of Daisies has asked you to make each petal equal 10. Can the children help? Prompt pupils by asking number bond questions such as, "What plus 0 equals 10?" "1 plus what makes 10?"

Dividing with Dennis the Dragon

Oral and Mental Warm Up

- Explain to the class that Dennis the Dragon has a problem. He has lost his dividing flames and cannot divide any more, and he wants the children to help him. Discuss the concept of division as sharing, and that sharing in two is the same as halving. Show the children the sign for division.

- Give each pair of children ten cubes. Ask them to divide the cubes into two equal groups so that each child has the same amount of cubes. Explain this is dividing by two. Repeat this activity using different numbers. What do the children notice about the numbers that are being used? They are all even numbers. Challenge the class to find out what happens if you use an odd number.

- Give each child a whiteboard. Write '12 ÷ 2 =' on the class board. Ask the children to write the correct answer. Repeat several times with different starting numbers.

- Make two sets of cards, one with dragons and even numbers written on, the other with flames and the corresponding sum when you divide by two. Give each child a card and when Dennis (you) 'ROARS', they find their partner!

- Children work in twos to make a division sum. Give each pair of children cards with a numeral, a division sign, an equal sign, a number two and the correct answer. For example, '6 ÷ 2 = 3'. Ask pairs to make their number sentence as quickly as possible.

- Play 'Guess the Number'. Show the children a number such as 8. Explain that this number has been divided by 2. What number do the children think you started with if the answer is 8? This could be extended to dividing by any number.

Focus of Learning

Understanding simple division

$8 ÷ 2 = ?$

Art and Display

- Use paint and collage materials to create a giant dragon for the display.

- Make and paint clay dragons.

- Draw, paint or use oil pastels to create individual dragon pictures. Use collage materials to make a scale effect.

- Make dragon puppets.

- Colour mix shades of red and yellow to make flames.

- Construct a giant whole-class dragon using recyclable materials or chicken wire and Modroc.

Practical Activities

- Produce a set of dragon division cards, with a dragon picture on one side and a division sum on the other. Prepare a set of gameboards with lots of flames and corresponding answers written on the flames. Ask the children to pick up a dragon card and cover the correct flame on their board. The object is to cover all the flames.

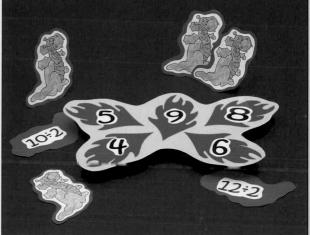

- Make a set of giant flames with numbers on and place them around the hall or playground area. Hand each child a division sum, asking them to work out and find the correct flame answer.

- Give each child in the group a whiteboard and different even numbers of cubes. Suggest they investigate how to divide the cubes into equal groups, writing the division sum on the whiteboard. For example, eight cubes could be '8 ÷ 2 = 4' or '8 ÷ 4 = 2', 12 cubes could be '12 ÷ 2 = 6', '12 ÷ 3 = 4' or '12 ÷ 6 = 2'.

- Play 'Divide with Dennis'. Make a set of different division sums. Cut the sums up into individual pieces. Back each piece of sum onto a different-coloured piece of card, for example all the division signs on red, all the equal signs on blue, all the answer cards on green, and so on. Each child picks five cards, one of each colour, and attempts to make a division sum. If successful, they keep the sum!

- Make a set of dragon cards with a variety of numbers, according to the ability of the children. Each child takes it in turn to pick two cards, for example 9 and 3, and tells you what the division calculation is. If an individual picks 10 and 4, they must explain why it is not a division sum.

Multiplication Madness

Focus of Learning

Beginning to use the multiples of 2

Oral and Mental Warm Up

- Count in twos while clapping hands together, clicking fingers and stamping feet. Count forwards and backwards to a given number. Start at different numbers.

- Play a counting fingers game. Ask the class to imagine that each finger is worth 2. Demonstrate counting in twos using fingers. For example, hold up three fingers and say, "This is worth six". Hold up four fingers, "This is worth eight", and so on. Ask the children to show a variety of multiples of 2.

- Show the children a 100-square with multiples of 2 highlighted in a different colour. Ask the children to tell you what they notice. Invite pupils to close their eyes. Remove a few of the multiples of 2 numbers and ask them to say what is missing.

- Ask the children to get into pairs and stand in a hoop. Together count how many groups of two there are and how many altogether.

Art and Display

- Paint or use collage materials to make a large bicycle for the display.

- Use collage materials to make cog patterns in multiples of 2.

- Use art straws to make bicycles.

- Use thread and hessian to sew bicycle pictures or wheel designs.

- Make a clay wheel. Ask the children to create a different texture in each part of the wheel by using clay tools or printing materials.

Practical Activities

- Practise problem-solving with bikes. Make a set of cards with a variety of problems written on relating to multiples of two. For example, 'How many wheels do eight bikes have?', 'There are seven bikes in the playground. How many wheels altogether?' 'A shop has three bikes. How many wheels is that?' 'One bike is sold. How many wheels are left?'

- Look carefully at a picture of bicycle and ask the children to draw either two pedals, two handlebars, two wheels, two lights, or any other multiple of two!

- Make sets of two using cubes. Write a multiplication sum on the board, for example '5 x 2'. Ask the children to show the answer with their cubes by sorting out five lots of two cubes.

- Make a set of hand cards showing multiples of 2 from 2 to 20, one set for each child. Instruct the children to place their hands over the hands on the card. Then write a multiplication sum on the board such as '10 x 2'. The children have to lift the correct finger so that the number 20 shows.

- Play 'Hoot the Hooter'. Stick a set of multiples of 2 cards on the board so that the numbers are not visible. Sit the children in groups with a whiteboard and pen each. On a hooter, one child from each group runs to take a card, returns to the group and shows the number. Each child has to write the repeated addition and the corresponding multiplication sum.

- Play 'Multiplication Mix and Match'. Place the cards face down. Each child picks three cards in turn. The object of the game is to collect a complete set. For example, '2 + 2', '2 x 2' and '4'. Each child can mix and match in turn until they have a complete set.

Funky Function Machines

Focus of Learning

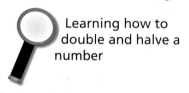

Learning how to double and halve a number

Oral and Mental Warm Up

- Discuss the concept of doubling. Stick two sets of large 1 to 10 numeral cards to the board. Invite the children to pick two numbers that are the same. Explain to the children that if they add them together, they have doubled the number.

- Give the children a whiteboard each and pick a number from the board. Write the number doubled on the whiteboard.

- Make a set of 1 to 10 number cards and a set with corresponding doubles. Give each child a card and ask the children to find their doubles partner.

- Draw on the board a large funky function machine to demonstrate the idea of doubling. In role as Daisy Double, put a number into the machine and challenge pupils to say what double number would 'pop' out from the other side. (Higher numbers can be used to extend more able children.)

- Discuss the concept of halving numbers. Give the children an even number of cubes and invite them to put them into two equal piles. Then hand out odd numbers of cubes and ask pupils to repeat the activity. What have they found out?

- On the board write a set of odd (1, 3, 5, 7, 9, 11, 13, 15, 17, 19) and a set of even numbers (2, 6, 10, 14, 18, 22, 26, 30, 34, 38). Challenge individuals to pick an even number and find its corresponding half.

Art and Display

- Use recyclable materials to make two large colourful function machines.

- Design and make funky function machines.

- Draw and paint funky function machines from junk materials.

Practical Activities

- Play 'Doubles/Halves Snap'. Have ready sets of 1 to 20 number cards and the corresponding number doubles. Children take it in turns to choose a card and shout "Double Snap!" or "Half Snap!" if they match the correct cards.

- Place the children's function machines around the room, with either a doubling or halving instruction on the front. Give a group of children a whiteboard, pen and a set of numbers. They visit each function machine, choose a number to put into it and write the answer on their whiteboard.

- Play 'Funky Function Machine'. Make a set of gameboards with doubling and halving function machines. Children look at the numbers going into the machine, writing the answer that pops out.

- Make two sets of cards, one set with funky function machines and an instruction to double or halve, and the other set with numbers 1 to 20 (or higher). Each child takes it in turn to turn over one function machine card and one number card. They either double or halve that number and record the answer on a whiteboard.

Pirate Pete's problems

Focus of Learning

Developing problem-solving skills

A pirate steals 7 treasure chests. He buries 3. How many are left?

Oral and Mental Warm Up

- Tell the children that Pirate Pete has lost all his pirates and he needs new 'shipmates' to help him steal treasure. Inform pupils that being a pirate is a very tricky and clever job, and that Pete needs to check if they are smart enough! Have ready a pirate's sack in which you have written cards with a variety of problems to solve. For example, 'There were 15 pirates cleaning the deck but three fell overboard. How many pirates were left?' Pick out one problem and ask the children to solve it.

- Play 'Problem Pandemonium'. Make a set of cards with pirate problems, similar to the one in the first activity, and a second set with the corresponding answers. Give out the cards and ask the children to find their partner.

Art and Display

- Create a large pirate ship using paint and collage materials.

- Paint and use collage materials to make individual pirates.

- Draw and design individual 'Jolly Roger' flags.

- Design and make a selection of pirate artefacts such as a pirate hat, eye patch and cutlass, using recyclable materials.

- Make pirate puppets.

Practical Activities

- Play 'Pirate Pandemonium'. In a large area each child wears their pirate hat and has a Jolly Roger flag to wave! They then sit on a boat (a small mat). Place a selection of large starred numbers around the room and hand each child a pirate problem to solve. On a given signal, the children 'swim' to find their answer star and bring this back to their boat as quickly as possible and wave their flag! To add excitement to the game, you could be a shark. If you touch the child on his or her return, he or she is out of the game.

- Introduce the children to the concept of two-step problems. For example, 'Four pirates are climbing the rigging, three more pirates climb up and two pirates fall off. How many pirates are left?' Ask the children to draw the problem and write the sum.

- Play 'Pirate Problem Bingo'. Make a set of treasure map cards with random numbers on, one for each child, and prepare a selection of corresponding pirate problem cards, written on sacks of gold. Each child takes it in turn to pick a sack and solve the problem. If they have the correct answer on their treasure map, they cover it with a counter or piece of gold!

Let's Go Shopping

- Turn the role-play area into a class shop. Ask the children to buy one sweet and give the correct money. Continue this activity by buying more than one sweet each time. Ask the children to total the amount spent.

- Give individuals an amount, for example 10p (you can vary the amount according to ability and experience). Ask them to spend exactly that amount at the shop and make a list of what they have bought. Then repeat this but ask the children to buy only one item. How much change will they receive?

Focus of Learning

Adding money and receiving change

- Play 'Shopping Spree'. Make two sets of cards: one set with shopping lists and prices (one card per child), and the other set with corresponding totals in purses. Give each child a shopping list card. They total their card and find the matching purse card.

Art and Display

- Create a shop front by painting a stripy canopy.
- Make giant sweets by using a variety of coloured paper.

Practical Activities

- Play 'Spin the Change'. The challenge of the game is to find how much change you get from a set amount, for example 10p (vary this amount according to ability). Instruct the children to sit in a circle. Place a spinner in the middle of a set of giant coins. Spin the spinner and ask the children to work out how much change from the set amount. Ask an individual to select the correct coins.

- Send a small group of children to the class shop, asking one child to be the shopkeeper. Prompt each child to buy two sweets. The shopkeeper adds the total. Each child checks this amount and gives the shopkeeper the correct money.

- Try the 'Let's Go Shopping' game. Make a simple shopping board game. Instructions on the board could include: 'Bought 3 sweets, spent 3p' 'Found 5p', and so on. Each child has 20p (as 1p, 2p, 5p and 10p coins). The object of the game is to move around the board and get to the finish without losing all the money. The child with the most money at the end of the game is the winner. You act as the banker and check the children's money each time.

- Play 'Shop 'til you Drop'. Make simple shop fronts (such as a shoe shop, grocer's, baker's, etc.) using A2 sheets of paper or card, suitable pictures and price labels. Place a selection of these shops around the classroom or school. Give each child 20p to spend. The object of this game is to spend exactly 20p by visiting all three shops. The children could write what they have bought on a large card purse.

- Repeat the above activity but with pupils spending exactly 50p at each shop they visit.

Money Monster

Oral and Mental Warm Up

- The money monster visits! Use a puppet as the money monster. He tells the children that he has taken all the money in the school but that he will only give it back if they can recognise the value of the coins he has in his bag! (Use giant coins.)

- Show some giant paper purses containing different amounts of money. Ask the children to add the money and find the total.

- Challenge the class to race to a £1. Split the children into two teams, A and B. Spread out plenty of large paper coins in front of a board (values 1p, 2p, 5p, 10p, 20p and 50p). One child from team A chooses a coin to sticky tack on one half of the board, then one child from team B chooses a coin and sticks it to the other half of the board. This continues until one team reaches exactly £1. Vary the amounts the children have to reach according to their ability.

- Try the 'Pass the Purse' game. Sit the children in a circle and pass around a purse containing some play coins while singing, "Pass the purse around and round, round and round, round and round. Pass the purse around and round. Stop right now" (to the tune of *London Bridge is Falling Down*). The child counts the money in the purse before the game is repeated. Change the amount in the purse each time.

- Play the 'Making Money' game. Give each child a giant coin. Then roll a die, which has amounts of money on each face (1p, 2p, 5p, 10p, 20p and 50p). The aim is for the children to get together into groups to make that amount. For example, if you roll 10p, two children who each have a 5p coin each could get together, or five children who have a 2p coin could get together.

Art and Display

- Draw and colour giant coins.

- Make money mobiles.

- Draw, paint and use collage materials to make a giant money monster.

- Draw, paint and use collage materials to create individual money monsters.

- Design and make a money monster using recyclable materials.

- Make coin rubbings on purse-shaped pieces of paper.

Practical Activities

- Try the 'Ten Pence' game. Make sets of 1p, 2p, 5p and 10p coin cards. In small groups, pupils take it in turns to pick a card. The object of the game is for each child to make 10p in total. This game can be played with any amount of money using different coin denominations.

We have drawn our own Money Monsters

We have found lots of ways of making different amounts of money

- Use the children's money monsters to make coin totals. Each child could choose his or her own total and find ways of making that amount.

- Play 'Money Pairs'. Make two sets of cards, one with coins on and the other with the written amount. The children take it in turns to turn over two cards. If they match, they keep the cards and the individual with the most cards at the end, wins.

- Try playing 'Money Hunt'. Make a set of sack-shaped cards with different coins and amounts drawn on. Vary the amounts according to the needs of the children. Place cards around the classroom. Give the children a whiteboard each and, in small groups, challenge them to find the sacks. When they find the sack, pupils total the amount and write it on their board or draw the correct coins for that amount.

- Play 'Money Lotto'. Make a set of gameboards with different values of coins, one board per child. Prepare a spinner with different monetary values written on (these must correspond with the values on the gameboards). The children take it in turns to spin the spinner and cover the appropriate sack with the correct coins.

- Try the 'Money, Money, Money' game. A large space is needed for this game. Split the class into six teams and give each team a different-coloured hoop. Place a large selection of giant coins on the floor in the middle of the room. Have a die prepared with different captions such as 'more than a pound', 'less than fifty pence', 'nearest to two pounds', 'more than five pounds', 'between sixty pence and one pound' and 'the same amount'. Play music such as *Money, Money, Money* by Abba while one child from each team puts a coin in the hoop. The children continue to do this until the music stops. You roll the die and read out the caption, and the team whose hoop is closest to what is written is the winner. Repeat the game.

Handa's Data Handling

Oral and Mental Warm Up

- Bring in a selection of different fruit to show the children. After tasting the fruit, make a class tally chart to show which fruit the class liked best.

- Make a human block graph by asking the children to stand in lines according to the fruit they liked best.

- Discuss with the class different ways of recording this information, for example as a pictogram, bar chart, block graph or pie chart.

- Use the information from the tally chart to create a giant class pictogram. Ask questions such as, "What is the total number of fruit altogether?" "How many more children liked pineapple than bananas?"

- Hold a class data quiz, with the children working in pairs and noting their answers on a whiteboard. Using the large class pictogram, ask questions about the chart for the children to answer after discussion with their partner.

Art and Display

- Paint and print a large basket for Handa's fruit. Draw and use pastels to create Handa's face.

- Use pastels to create the different fruits.

- Use watercolour pencils to make pineapples.

- Create a clay basket of clay fruit.

- Use hessian and thread to create 3D fabric fruit.

- Make observational drawings of the fruits in the story.

Practical Activities

- Ask the children to use the knowledge they have acquired about graphs to do a similar activity in another class, perhaps tallying up favourite colours.

- Encourage the class to record the information from the tally chart as a pictogram. Then invite them to record the information in at least two more ways.

- Ask the children to think of five questions to ask a friend about the information on their own graph.

- Use cubes, beads and wooden blocks to make a 3D block graph to show the results of the fruit-tasting session (see page 46).

Mr Triangle

Oral and Mental Warm Up

- Dress up as a Queen or King of 2D Shapes with lots of 2D shapes stuck to you (square, circle, rectangle). Pointing to each shape in turn, prompt the class to identify it and to name its properties.

- Play musical 'Pass the Shape'. Ask the children to sit in a circle and pass 2D shapes around. When the music stops, the child stands up and names the shape and one of its properties.

- Play 'Guess the Shape'. Hide most of a shape behind a piece of cloth or card and challenge the children to identify it. Keep showing a bit more of the shape until the class have identified it.

- Put some 2D shapes in a feely bag, and invite individuals to feel one without looking and to identify it.

- Suggest one child leaves the room and ask the rest of the class agree on one shape. The child comes back in the room and tries to guess the shape by asking questions. The rest of the class can only reply by answering 'yes' or 'no'.

Art and Display

- Paint giant 2D shapes in bright colours for display. Make a collage face for each shape.

- Make shape pictures using 2D shapes.

- Design and make shape puppets using the shapes in the story.

- Make a class clay shape picture. Ask each child to create a shape out of clay. Place each shape on a large board for display.

- Print repeating patterns using different shapes.

Focus of Learning

Developing an understanding of 2D shapes

Practical Activities

- Play 'Shape Treasure Hunt'. Place 2D shapes around the classroom. Give each child a card with a shape property written on it and challenge pupils to find the matching shape.

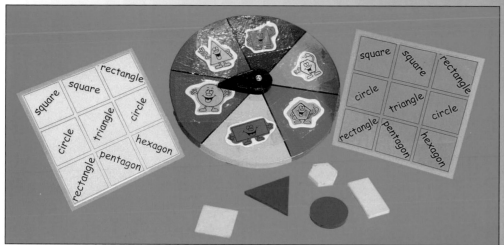

- Try the 'Spin a Shape' game. Prepare a set of spinners with pictures of shapes written on and a gameboard with names of shapes. Ask the children to spin the spinner and find the corresponding shape.

- Involve the class in 'Shape Snap'. Make a set of cards with shape pictures on and a set with properties. The children then match the two, shouting 'Shape snap!'

- Sort a variety of 2D shapes according to their properties by putting them into a hoop Venn diagram.

- Give half the class a shape each and the other half a card with the shape's property written on. Ask the children to find their partner as quickly as they can.

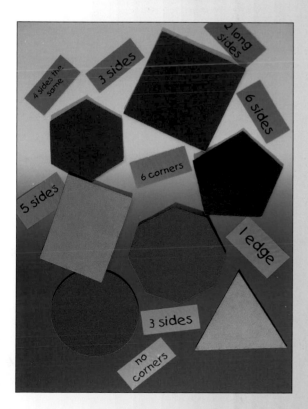

- Play 'I Spy a Shape' around the classroom. Use everyday objects such as a clock, book and table. Ask a child to choose an object and give its properties. For example, "I spy with my little eye something that has no corners." (Answer – a clock) If the class does not guess the object first time, the child must give another clue.

Treasure Map

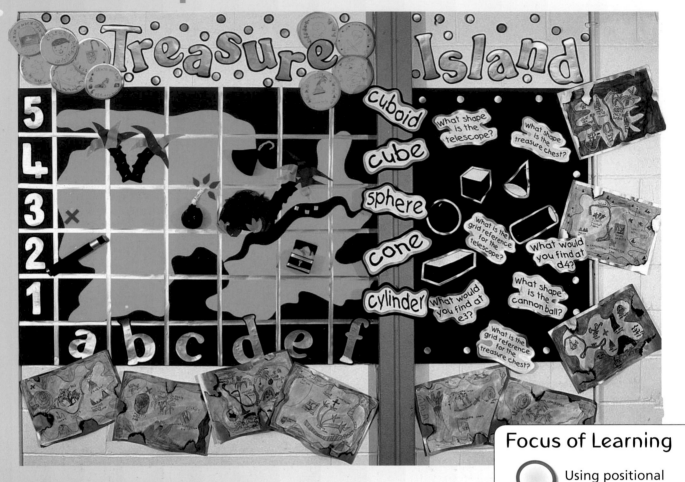

Focus of Learning

Using positional language and the names of 3D shapes

Oral and Mental Warm Up

- Prepare a treasure map with places named after 3D shapes such as sphere, cube, prism and so on. Also have ready a booty bag containing 3D shapes. Empty the bag and ask the children to match the shapes with the names on the map.

- Prepare some cards with the shape properties and names. Look closely at each shape with the class and discuss the properties of each one. Invite the children to help you match the name of the shape, the 3D object and a property clue. For example, give the clue 'no corners'. The children should then pick out the sphere and point to the sphere on the map. (They may need extra clues for some shapes.)

- Play 'Guess my Shape'. Ask a child to hide a 3D shape in a treasure chest (decorated box). One child picks out a shape and gives clues to the rest of the class for them to guess what it is.

- Talk about co-ordinates – what they are and how they work.

Art and Display

- Make a large treasure map for display.

- Make pirate hats, eye patches, belts and treasure using recyclable 3D materials.

- Draw and colour individual treasure maps. Use tea- or coffee-stained paper for the background and burn the edges to create an ancient effect.

- Make clay or Modroc treasure islands.

Practical Activities

- Place a selection of 3D shapes around the classroom, hall or outside area and use the property cards from the 'Oral and Mental Warm Up'. Each child sits in a hoop. Pick a clue from the treasure chest and, on the signal 'Find the Booty', the children rush to find the correct shape and bring it back to their hoop.

- Play 'Steal the Booty'. Make a set of treasure island maps with co-ordinates but no treasure. Also prepare several small sets of 3D pirate treasure such as a sphere (cannon ball) and a cuboid (bar of gold). The children work in pairs with a barrier between them so that they cannot see each other's board. Each child places his or her 3D treasure on their treasure island. The object of the game is to steal the other person's treasure by guessing the correct co-ordinates. For example, child A says, "Is your cuboid bar of gold in B2?" If correct, child A steals the booty!

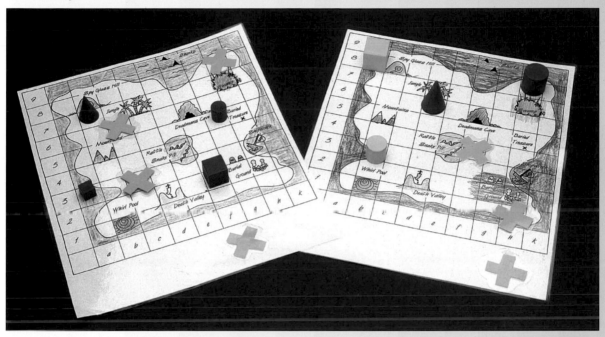

- Play 'Spin for Gold'. Ask the children to sit in a circle. Place the 3D treasure in the middle with a spinner, which has the names of the shapes written on. Child 1 spins the spinner and collects the shape that it lands on. Child 2 spins the spinner and collects the shape. If a child lands on the same shape as a previous child, that child still takes the shape. When all the children have had a turn, the ones with a piece of 3D treasure collect a gold coin (a large paper coin). Repeat the game several times. At the end of the session, the child with the most gold coins is the winner.

- Play 'I Spy Treasure'. Place the treasure (3D shapes) on a large map, either painted on the playground or made out of paper. Ask the children to say where each piece of 3D treasure is by naming the co-ordinates.

- Then play 'Give us a Clue'. Invite a child to pick a clue out of a treasure sack, for example 'I have 6 faces'. The children write the correct co-ordinate of the shape (as placed in the previous activity) on a whiteboard.

Royal Right Angles

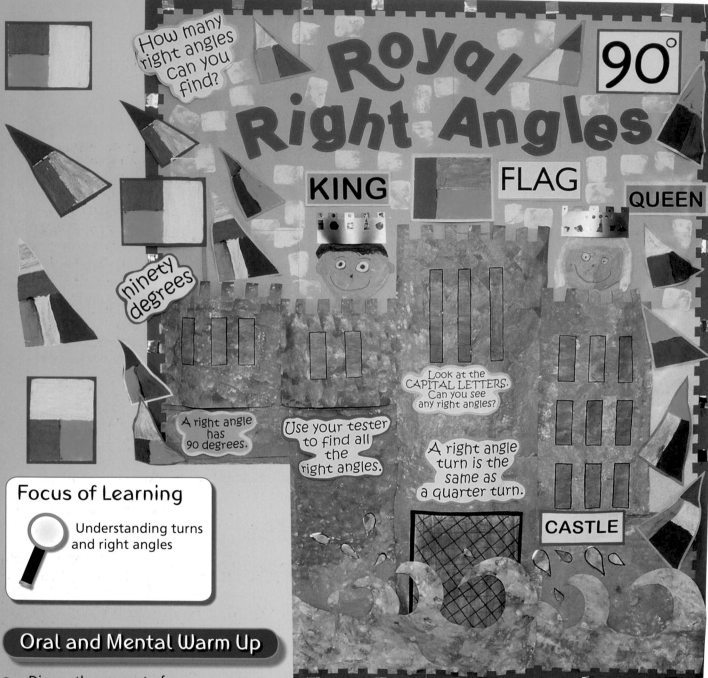

How many right angles can you find?

Royal Right Angles

90°

KING

FLAG

QUEEN

ninety degrees

A right angle has 90 degrees.

Use your tester to find all the right angles.

Look at the CAPITAL LETTERS. Can you see any right angles?

A right angle turn is the same as a quarter turn.

CASTLE

Focus of Learning

Understanding turns and right angles

Oral and Mental Warm Up

- Discuss the concept of turning with the children. Ask them to stand up and make a whole turn, then a half turn, then a quarter turn.
- Play 'Recognise the Right Angle'. Put a selection of 2D shapes and pictures on the board, some of which have obvious right angles. Ask the children to spot the right angles.

Art and Display

- Paint a giant 2D right-angled castle. Use sponge print to create a brick effect. Paint right-angled flags. Create a king and queen for display using collage materials.
- Make clay castles.
- Produce right-angled crowns using card and collage materials.

Practical Activities

- Make a right-angle tester by folding a piece of paper into quarters. Ask the children to use their tester to find right angles on the display castle and around the classroom.

- Work in small groups on the floor to create a right-angle maze for a programmable toy to get to the 3D castle. Record the moves the children think the toy will make. For example, 'F2 (forward 2) RT 90 (right-angle turn) F3 (forward 3)'.

- Play 'Spot the Right Angle'. Make a set of simple pictures of castles, kings or queens, all out of right angles. Make a die with the word 'right angle' on four sides and a star on each of the two remaining sides. Pupils take it in turns to roll the die. If it lands on the word 'right angle', the children mark a right angle on the picture. If they land on a star, they miss a go.

- Use geoboards and elastic bands to make right-angle shapes. Extend the children's learning by asking them specific questions such as, "Can you make a pentagon with two right angles?" "Can you make a hexagon with four right angles?"

- Play 'Royal Right Angles'. Draw a simple 2D castle. Draw a line down the middle and label one side 'Right Angles' and the other side 'Not Right Angles'. Make a die with the shapes square, circle, hexagon, pentagon, right-angled triangle and an equilateral triangle. Also have ready card exam ples of the same shapes, at least four of each. The children take it in turns to roll the die, pick up the appropriate shape and place it on the correct side of the castle.

- Play 'Right Angle Snap'. Make a set of cards, some with right-angle shapes and some without.

Sir Symmetry

Oral and Mental Warm Up

- Dress up as Sir Symmetry and ask the children "What is symmetry?" "Can you see anything symmetrical in the classroom?"

- Provide a sack of ten or so cardboard shields, some of which have symmetrical patterns and some of which don't. Invite the children to sort them out.

- Help Sir Symmetry to design a new school shield. Make a large blank shield out of card and cut out a variety of shapes and symbols. Ask each child in turn to place one of these on the shield to make a symmetrical pattern.

- Discuss the concept of 'line of symmetry'. Give pupils some paper shapes, each of which has one or more line of symmetry. Challenge the class to fold the shapes along lines of symmetry. How many lines of symmetry does each shape have?

- Make a table display of symmetry words and symmetrical shields.

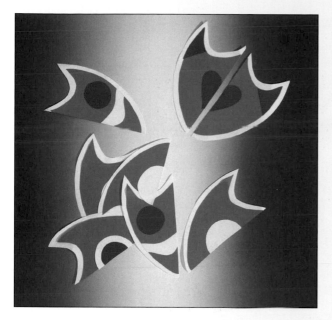

Art and Display

● Use paint and collage materials to create a giant symmetrical knight.

● Use a computer paint package to draw a symmetrical knight or shield.

● Use binca material and thread to create a symmetrical pattern on a binca shield.

Practical Activities

● Play 'Share a Shield'. Make a set of large blank shields and cut out a variety of brightly-coloured paper shapes. Also create a set of shape dice. Children play in pairs. Give each pair a shield, some shapes and a die. The children take it in turns to roll the die, pick up the correct shape and place it on the shield. The object of the game is to make a symmetrical pattern on the shield. (Alternatively, this game can be played by asking the children to draw the shapes on the shield.)

● Try 'Search for the Shield'. Make a set of symmetrical shields, cut them in half and place facedown on the table. The children take it in turns to pick up two halves. If they match, they keep the shield.

● Use the halves of shields from 'Search for a Shield'. Give each child a half-shield, a piece of paper and a mirror. Ask the class to place the mirror on the line of symmetry to complete the shield. The children draw the other half of the shield. As an extension, pupils could draw their own half-patterns and ask a partner to complete it.

● Give each child a shield cut out of squared paper. Ask them to draw the line of symmetry and complete a pattern on one side of the shield. Give the shield to a friend to complete.

Funny Bones

Focus of Learning

Using different units of measurement for length

Oral and Mental Warm Up

- Discuss and compare the heights of the children in the class. Talk about the tallest and shortest.

- Either you or a child can dress as a skeleton with removable bones. Ask the children to find the longest and shortest bones.

- Hand out a selection of paper bones. Challenge the class to order the bones from the shortest to the longest.

- Invite the children to find three items or more around the classroom that are longer/shorter than selected bones.

- Estimate the length of the bones using non-standard and then standard units of measure.

- Ask the class how they could measure different parts of the body. For example, the length of their thigh bone, around their wrist or skull. Ask pupils to work in pairs to estimate, measure and record their findings.

Art and Display

- Make large 'funny bones' characters for display based on the characters in *Funny bones* by Janet and Allan Ahlberg (published by Puffin Books).

- Make individual skeletons using art straws.

- Use a computer paint package to draw simple skeletons.

- Ask the children to look carefully at a large image of a skeleton. Using their observational skills, ask them to sketch either the whole skeleton or a selected part.

Practical Activities

- Using the large 'funny bones' characters from the display, ask the children to estimate and measure the length of the different bones in either non-standard or standard units of measure, according to children's understanding, and to record their results using a whiteboard.

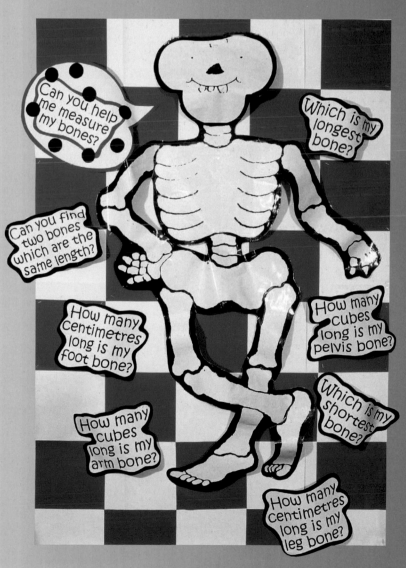

- Hold a bone hunt. Give the children paper bones of different sizes and prompt them to find things around the classroom that are longer than, shorter than or the same size.

- Make some cardboard thigh bones of differing lengths using multiples of 10cm. Make some smaller bones of multiples of 5cm. Ask the children to estimate how many 5cm bones would you need to make a 30cm bone, how many 10cm bones would you need to make a metre, and so on.

- Make a 'funny bones character' one metre long. Ask the children if they can jump a metre. Then challenge the children to search the classroom to find things that are one metre long.

Santa's Parcels

Oral and Mental Warm Up

- Prepare a set of Santa parcels, all of different sizes and weights but in pairs of weights that will balance, such as one at 50g and two at 25g; one at 1kg, one at 750g and one at 250g. Dressed in role as Santa, show the class the parcels and a set of balance scales. Ask them to weigh the parcels and sort them in order from the heaviest to the lightest.

- Ask the children to guess which parcels might balance each other. Weigh the parcels to check.

- Play 'Guess My Weight'. Use a range of non-standard units to weigh a selection of Santa's parcels, such as marbles. Ask the children to estimate first and record their guess on a whiteboard.

- Play 'I-Spy a Kilogram Parcel'. Discuss with the class the need to use standard units of measurement. Introduce a kilogram weight and pass it around the circle so that the children experience what a kilogram feels like. Ask individuals to pick a parcel that is about the same weight as a kilogram.

- Play 'Spin the Bottle'. Sit the children in a circle and place a selection of parcels in the centre of the circle. Spin the bottle and whichever child it points to has to pick up a parcel that weighs about a kilogram. Check how close they are using the scales.

Focus of Learning

Using weight measurements

Art and Display

- Paint and collage a large Santa on his sleigh.

- Design and print wrapping paper for Santa's parcels.

- Design and make sleighs using recyclable materials. Spray them red and gold.

- Paint or collage giant toys for Santa's sleigh.

- Print a snowflake background using the edge of a piece of card. Ask children to paint a picture of Santa and his sleigh.

- Make stars using triangles or pentangles to use on the display.

- Provide a selection of small pieces of simple Christmas wrapping paper. Ask the children to select one piece and to glue it on to a larger piece of white paper. Using watercolour pencils, ask the children to continue the pattern.

Practical Activities

- Turn a set of balance scales into Santa's sleigh by putting a picture of his sleigh on the front. Ask the children to balance a selection of parcels to find out which parcels are heavier or lighter. Ask the class to guess first.

- Give the children a parcel that weighs exactly one kilogram. Ask them to find as many objects around the classroom that weigh the same as they can.

- Play 'Balance Santa's Sleigh'. Make a set of brightly-coloured sleighs with a weight written on the front. Then prepare a set of cards with parcels and weights on. Each child picks a card in turn and puts it on a sleigh. The object of the game is to balance the sleigh with the correct weight of parcels.

- Play 'Snap the Parcel'. Make a set of cards with a picture of a sleigh and a weight in kilograms. Also create a set of cards with a number of parcels that total a corresponding weight. For example, a sleigh card could have 50kg written on it and a child would turn over a parcel card that might have five parcels each weighing 10kg. Snap!

Estimating Eggs

Oral and Mental Warm Up

- In role as Mother Hen, explain to the class that you have forgotten how many eggs you have laid. Place a number of balls in a basket and ask the class to estimate how many eggs there are.

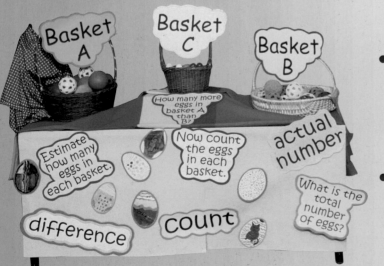

Focus of Learning

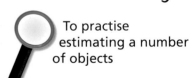

To practise estimating a number of objects

- Now show seven different-sized baskets – one for each day of the week. Challenge the children to estimate how many eggs Mother Hen (you) will have to lay to fill each basket. The children write their estimate on a whiteboard. Then ask a child to help you fill each basket and count the actual number.

- Ask the children to sit in a circle. Give each child a number card from 1 to 30. Place a selection of 'eggs' in the middle of the circle. Ask the children to estimate the number of eggs and to stand up if they think they have the correct card. Count the eggs to see who is correct.

Art and Display

- Use paint and paper collage to make a giant hen for the display.

- Use papier-mâché to make giant eggs. Paint a colourful design on each one.

- Employ marbling technique to create patterns on egg-shaped paper.

- Try oil pastels to make a colourful design on egg-shaped paper.

- Decorate hard-boiled eggs using poster paints.

- Use recyclable material to design and make a basket to carry the eggs.

- Make clay egg cups.

Practical Activities

- Play 'Eggs in the Egg Cups'. Place a number of egg cups on the table, say eight, and put a basket of ten eggs (balls) on the table. Ask the children to estimate if there are enough egg cups for each egg. Are there too many or too few? (Any numbers can be used according to the ability of the children.)

- Play 'Estimating Eggs'. Make a set of large eggs with spots on (using numbers up to 20) and provide a number line. Ask the children to pick an egg and estimate how many spots it has. Give pupils a 0 to 20 number line and ask them to record their estimate on it. The children count the spots and record the actual number of spots, before finding the difference between the estimate and the actual number.

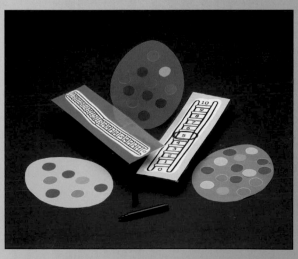

- Make a spinner with multiples of 10 on it and have ready a set of spotted egg-shaped cards – the number of spots varying between 5 and 100. The children take it in turns to spin the spinner. They estimate which egg card has the number of spots nearest to the number on the spinner. They pick up the card and count the spots. The children in the group discuss and decide if it was a near estimate.

What's the Time, Mr Wolf?

Oral and Mental Warm Up

- Show the children a clock. Make a collection of different types of clock.

- Make the parts for a brightly-coloured clock face: a large circle, numbers 1 to 12 and two hands. Give out the pieces to each child. Play some gentle music and ask the children to put the pieces of the clock together starting with the large circle.

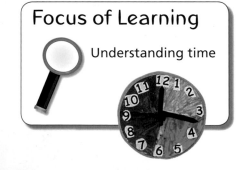

- Ask the children to describe their day, for example, "I got up at 8 o'clock." Show the time on the clock face. Relate analogue clocks to digital clocks.

- Make a set of cards showing matching analogue and digital clock faces and times. Begin with 'o'clock' times, before introducing other times according to the needs of the children. Give each child a time card and ask them to find their analogue or digital partner.

- Explain the concept of one hour later and one hour earlier. Give the children a time and ask them to show one hour later and one hour earlier on their clocks.

Art and Display

- Paint or use collage materials to make large clocks for the display.

- Paint a giant Mr Wolf.

- Design and make individual clocks.

- Make wolf masks to use for playing What's the Time, Mr Wolf?

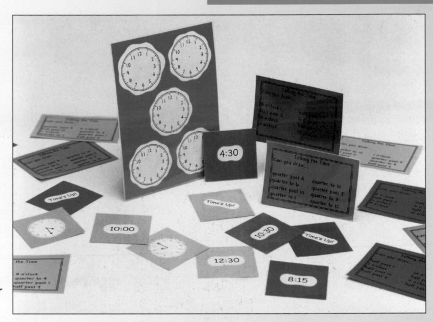

Practical Activities

- Try some timed activities using a variety of timers. Ask the children to estimate how many times they can jump, hop, write their name, or how many cubes they can fit together in a set time. Set the timer and have a go!

- Play 'Match the Time'. Make a set of cards with matching analogue and digital times and play Pelmanism by putting the cards face down on the table. The children take it in turns to pick up two cards. If the times match, the children keep the cards.

- Children practise drawing clocks and writing the 12 numbers in the correct places. Extend this activity by using a one-minute timer.

- Play 'Make a Clock'. You will need a 1 to 12 die. Give each child in the group a clock face, clock hands and the numbers 1 to 12. They roll the die and whatever number it lands on, the children stick the number in the correct place on the card.

- Play 'What's the Time, Mr Wolf?' Make a large clock with movable hands on one side and a knife and fork on the other side. Dress up as Mr Wolf. The children walk behind you shouting, "What's the time, Mr Wolf?" Show a time on the large clock and the children have to shout the correct time. When the knife and fork is shown, you shout, "It's dinnertime!" and chase the children.

- Play 'Three Times'. Make a set of cards showing different times, allowing one card per child. On one card show, for example, ten o'clock. On the second card show one hour earlier and on the third card show one hour later. Hand out the cards to the class. When you shout the words, "Three times", the children get in to the correct group of three.

Monday's Child

Focus of Learning

Ordering and remembering the days of the week

Oral and Mental Warm Up

- Read the poem 'Monday's Child is Red and Spotty' by Colin McNaughton in *There's An Awful Lot of Weirdos in Our Neighbourhood* (published by Walker Books, 2000).

- Prepare a set of pictures relating to the poem and make a set of days of the week cards. Give the cards to the children and ask them to arrange themselves in the correct order.

- Hand each child a card with a day of the week written on it. Ask seven children to stand in a circle. Ask questions such as, "What day comes before Wednesday?" or "Two days after Thursday?" The child holding the correct card sits down.

- Play 'What's Missing?' Give each child a whiteboard. Then you say the days of the week in order but missing one out. Pupils write the missing day on their whiteboard.

- Make four sets of days of the week cards, each on a different colour card (for example one set in red, one in blue, one in yellow and one in green). Give each child a card. On the sound of a hooter or bell, the children quickly get into colour groups and order themselves correctly.

Art and Display

- Paint and use collage materials to make a giant mother figure.

- Paint and use collage materials to make the children in the rhyme.

- Ask the children to find out on what day they were born. Suggest they draw or paint a picture of themselves using the rhyme in the poem, for example a child born on Monday would draw him- or herself red and spotty.

- Help the children to cut out letters from a newspaper to spell the days of the week. Arrange them as a collage.

- Write one of the days of the week using wax crayons. Paint a wash of colour over the wax crayon.

Practical Activities

- Play 'Spin the Week'. Make a gameboard showing the children from the McNaughton poem and prepare a 'days of the week' spinner. In turn, ask pupils to spin the spinner and, whatever day it lands on, to cover the appropriate picture.

- Play the 'Ladder' game. Give each child a picture of a ladder and have ready a set of days of the week cards for each child in the group. Put the cards in a bag and ask each child to pick a card out of the bag and put it on the ladder. When each child has collected all seven days, turn a timer over and ask them to order the days correctly.

- Make a 'days of the week' gameboard for each child in the group and a set of cards with clues such as 'the day between Thursday and Saturday', 'the day after Sunday', and so on. Each child takes it in turn to pick a clue and cover the correct day on his or her board.

- Play 'Hunt the Days of the Week'. Make a set of large pictures from the poem and place them around the classroom or school. Give each child a whiteboard. Ask pupils to hunt the pictures in the correct order. When they locate a picture, ask them to write the correct day three times on the whiteboard. To help the children, the correct day could be written on the back of each picture.

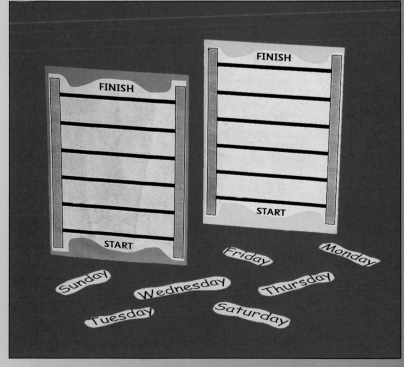

Calendar of Cars

Oral and Mental Warm Up

- Sit the children in a circle on the carpet. Chant the months of the year, asking the children to stand up when their birthday month is said.

- Use a jingle ball. Roll it to a child across the circle and say a month in order. Repeat for all the months.

- Try the 'Guess the Month' game. Collect items to represent each month, for example Christmas card (December), red heart (February), witch's hat (October), etc. Children should try to identify the corresponding month.

- Write a set of months of the year on large colourful labels. Using the items from 'Guess the Month', ask the class to match the month with the item. This game can be extended by removing an item and a label and challenging the children to identify which month is missing.

- Play 'Calendar Count Down'. Give 12 children a month of the year card each. The rest of the class count down from 20 while the 12 children get into the correct order. Repeat until every child has had a turn. The object of the game is to beat the 20 count down.

Art and Display

- Draw and paint large colourful cars. Ask the children to personalise their car by decorating it with a pattern.

- Use pastels to draw faces. Add collage materials such as wool for hair.

- Sketch and label parts of a car.

- Use brightly-coloured paper to create a picture of a car.

- Make a crazy colourful car using recyclable materials.

66

Practical Activities

• Play 'I Spy My Birthday Month'. Place large 'months of the year' labels around the room. On a given signal, ask the children to stand beside their correct birthday month. Count how many pupils have birthdays in each month and note the results on the board.

• Use the data collected from 'I Spy My Birthday Month' to draw or make a pictogram, bar chart or block graph. Ask questions about the data such as, "How many of you have a birthday in December?" "In which month do most of you have your birthday?" "How many more of you have birthdays in January than June?" Then invite the children to create their own questions by using the above data.

• Play 'Birthday Pairs'. Make two sets of colourful cars, one set with ordinal numbers from first to twelfth written on and one set with the months of the year written on. Each child takes it in turn to pick one card from each set. If they match, the child keeps the cards. For example, 3rd and March, 6th and June and so on. The winner is the child who collects the most cards.

• Play 'Race to Your Birthday'. Make a giant racetrack and mark it with ordinal numbers up to the twelfth. Using the months of the year cars from 'Birthday Pairs', ask the children to take it in turns to pick a car and place it next to the correct ordinal number.

• Play 'I Spy Calendar Clues'. Place some months of the year clues around the classroom or school and give each child a whiteboard and pen. Challenge pupils to find each clue and write down the corresponding month.

Captain Capacity

Oral and Mental Warm Up

- Tell the class that Captain Capacity is going to visit the classroom to explain about capacity. In role as the Captain, talk about the meaning of the word 'capacity' and show them the word written on the board. Then teach the class a capacity jingle to the tune of *London Bridge is Falling Down*:

  Captain Capacity is filling his bath, filling his bath …

  Tell him now!

  Captain Capacity wants you to guess, wants you to guess …

  How many is it?

  Pour the buckets in the bath, in the bath …

  How many buckets full?

As the children sing, Captain Capacity fills a bucket, mug, hot-water bottle, egg cup, etc. with water.

- Have ready a selection of containers holding different amounts of water so that Captain Capacity can ask the children which holds less, more and the same.

- Show the class a litre bottle and ask them to estimate how many litres would fill the Captain's bucket, mug, bath, egg cup, etc.

Art and Display

- Make a large bath for display.

- Create bubble prints using thin paint and washing-up liquid. The children blow through a straw to create lots of bubbles and make a print by placing a piece of paper on top.

- Make a small clay chamber pot.

- Design a pattern on chamber-pot-shaped paper.

- Use stencils to create a design on bath-shaped paper.

a mug

your bath !

an egg cup

a pan

How many litres to fill ....

?

1 litre

a bowl

Capacity

about 1 litre

less than 1 litre

more than 1 litre

Practical Activities

- Provide the children with a variety of different-sized containers and measures. Ask them to estimate how many buckets of water it would take to fill a bath, how many mugs it would take to fill a bucket, how many egg cups it would take to fill a saucepan. Allow pupils to test their predictions.

- Ask the children to find things that contain more than a litre, less than a litre, the same as a litre.

- Play 'Captain Capacity's Container' game. Each child chooses a container. Make a spinner with different measures written on such as spoon, egg cup, bowl, mug and saucepan. Each child takes it in turn to spin the spinner. Whichever measure they land on they add that amount of water to their container. The object of the game is for each player to fill their container exactly to the top without it overflowing. If their container overflows, they are out of the game!

- Provide the children with a variety of different containers. Challenge them to order and label the containers according to how many litres they might hold. Invite pupils to test their predictions and record the results on a block graph.

Numeracy Resources

As teachers we have found these resources to be an essential part of everyday maths in the primary classroom. These resources must be BIG, BRIGHT and BEAUTIFUL to ensure that we are accessing children's visual learning. All of the resources can be used as a constant teaching aid and provide interactive learning for the children. If they are on permanent display they not only brighten up the classroom but the children will absorb the knowledge subconsciously and use it to extend and develop their own individual learning.

For example:

- mathematical vocabulary
- giant colourful numbers and words
- number sequences/number patterns
- 100-square
- multiplication 100-square
- terrific times tables tree
- giant money
- giant shapes.

Resources for all occasions:

- feely box
- blank bingo gameboards
- beanstalk gameboards
- giant blank wooden spinners

- big digit cards
- number grid
- blank caterpillars/snakes
- giant dice

These resources are invaluable because, as the heading suggests, they can be used for a wide variety of activities. Our book is about a very practical, fun approach to mathematical learning, which requires a variety of stimulating resources. We know as well as anyone that it takes time and energy to make resources, but we are convinced that it is time well spent – and here is how to manage the workload:

- Make a general bank of resources. Never make just one.

- Use ideas that can be adapted for other skills. For example, the blank wooden spinners can be used for shape games, number games or money games simply by changing the pictures on the spinner by using tacky back or Blu-Tack. Children love to spin!

- Laminate everything to make it hardwearing and durable.

- Have a 'Wonderful Workshop' day each term when all parent helpers, classroom assistants and student teachers join together to make the resources. (One master will have already been made by you for them to copy.)

- Each week, try to make one new game or resource. This will spread the workload and make it much more manageable.

Resources go Outdoors

It is essential that the outside learning area is an extension of the classroom. Therefore there must be opportunities for the children to develop and extend their mathematical thinking and understanding. By creating stimulating mathematical resources, pupils will have the opportunity to consolidate their learning and see maths as part of their everyday life and not something that just happens for one hour a day in the classroom. For example:

- 100-square
- Snakes and Ladders gameboard
- co-ordinates gameboard
- numbers
- clock
- hopscotch
- number snake

Resources Really Help

It is essential for teachers to develop the children's independent learning. In this book we have suggested many games and activities that need the support of an adult. There are many times when pupils need to be encouraged to work independently. By providing them with a range of resources to support their learning, they will become confident independent learners. For example:

- 100-squares
- multiplication squares
- number ladders
- hand cards
- coin cards
- number tiles
- number staircases
- number card.

Captain Capacity (page 68)